Construction QA/QC Systems that Work:

Case Studies

Proceedings of a session sponsored by
the Construction Division of the
American Society of Civil Engineers
in conjunction with the ASCE Convention
in Denver, Colorado

May 1, 1985

Edited by George Stukhart

Published by the
American Society of Civil Engineers
345 East 47th Street
New York, New York 10017-2398

The Society is not responsible for any statements
made or opinions expressed in its publications.

Copyright © 1985 by the American Society of Civil Engineers,
All Rights Reserved.
Library of Congress Catalog Card No.: 85-70500
ISBN 0-87262-442-0
Manufactured in the United States of America.

CONTENTS

INTRODUCTION

Quality Assurance / Quality Control: Case Studies
George Stukhart 1

COMMENTARY

Quality in the Constructed Project: A Workshop Report
C. B. Tatum 5

SESSION PAPERS

The Corps Blue Ribbon Panel
Homer Johnstone 15
QA on the Texas Commerce Tower
Paul Little 21
Dallas North Tollway Project
Stephen B. Quinn 26
Quality in the Work Place
Howard A. Peek and Donald Brown 38

Subject Index 43

Author Index 44

QUALITY ASSURANCE/ QUALITY CONTROL: CASE STUDIES

INTRODUCTION

George Stukhart*, M. ASCE

A number of articles have recently appeared in Civil Engineering(1,2,5,6) and the ASCE journals (3,7), indicating that there is considerable interest in the management of quality in the construction process. Most of this attention has been directed at the role of the engineer as designer and inspector in construction. There has not been much discussion of quality management by the contractor. Perhaps engineers have the opinion of some writers (1,4), that the contractor will not respond to quality requirements unless there is a strong inspection program.

The papers in this session will hopefully illuminate the issue of responsibility for QA/QC by describing actual situations where quality work is being done in the field. The recent article in Civil Engineering (2) illustrated that there are many such methods, and these can be tailored to fit the needs and capabilities of the owner, the technical nature of the project, and the contractor's qualifications. The papers that follow are only a small sample of the many ways quality can be achieved. A deliberate attempt has been made to endorse efforts by Contractors to initiate quality programs without the stimulus of regulation.

The key to any quality program is strong management support. The Corps of Engineers study highlighted the importance of senior and intermediate level emphasis on quality. The Corps has long adhered to the doctrine that the Contractor has an important role to play in quality; their study, reported by Homer Johnstone, chairman of the Blue Ribbon Panel on quality assurance/ quality control, shows how important management commitment and education are to keeping everyone quality conscious. As part of the educational process, field people must believe that they are involved in achieving quality.

The next paper, by Stephen Quinn, shows how a designer is being used to effectively enforce quality on a major civil project. HNTB serves as GEC and Construction Manager

---

 Associate Professor, Civil Engineering Department, Texas A&M University, College Station, Texas 77843.

for the Dallas North Tollway. The advantages of being knowledgeable in the design and construction of the work are evident in the paper. Perhaps the most significant attribute of this project is the training program for inspectors. This type of training not only provides technical proficiency; it also serves to make everyone quality conscious at the start of the project.

A contractor quality program, which is initiated because the contractor believes in its cost effectiveness, is the most desirable solution to construction quality. Such a program relies on the initiative of field people who are trained and motivated. The more people involved in thinking about quality, the better. Paul Little of Turner Construction relates his experiences in making quality work on the Texas Commerce building, a very significant challenge. This is truly a case study of a quality control system that works for many of the same reasons as above: knowledgeable, dedicated people, prevention of problems, managerial commitment to quality, and a good attitude. Mr Little's article is excellent for the specifics of his own personal learning experience. Everyone can gain by reading it.

Howard Peek and Donald Brown of Brown and Root introduce what may be the future: Zero defects and quality motivation at the worker's level. There is no scarcity of literature on these subjects but very few have gone as far as trying a long-range program in their own companies. The authors are to be commended for making the effort. Again the emphasis is on the human side, but, interestingly, the authors relate quality to planning. Their theme is that prevention of defective work is a function of planning, motivation and training. They recognize that these are all necessary to improve work methods.

"While recognizing that it is not justified to spend a great deal of time and money to eliminate minor faults entirely, it must be agreed that a target of minimum (if not zero) faults is much easier to put over with workers than one of minimum cost.(3)."

The session moderator's own experience working with Owens-Corning Fiberglas(2) has reinforced his belief that senior management must give wholehearted support to quality goals. The crafts and their immediate supervisors must believe in the program to make it work. Quality planning and attitude have the greatest influence on the success of any such program. The OCF program has had 100% support from top management, and the contractors have generally supported the program, despite the detailed nature of the checklists and the extra time involved.

This session is a logical follow-on to the ASCE

workshop on <u>Quality in the Constructed Project</u> held in November 1984. Since there are many issues that are common to that workshop and to this session, C.B. (Bob) Tatum will summarize their findings. He has provided a full report which is included in the proceedings of this session. He emphasizes several requirements for quality management:

   a. The need for a knowledgeable field representative, preferably the resident engineer. This person must be familiar with the design and can really influence construction quality.

   b. Construction quality must be built in, rather than inspected in the final product.

   c. Quality must be achieved within the constraints of the project.

   d. People in the field need a greater understanding of requirements, expectations, methods and criteria. This means that codes and standards are not always available nor fully understood.

   A great deal of this field commitment can be achieved through realistic, accurate plans and specifications, something most engineers can influence. The best construction quality program will soon flounder if the field has no confidence in the plans and specifications.

REFERENCES:

1. Casper, Kenneth L., "Limited Field Inspection vs. Public Safety", <u>Civil Engineering</u>, Vol. 54, No. 5, May 1984, pp 52-55.

2. Fairweather, Virginia, "The Pursuit of Quality: QA/QC", <u>Civil Engineering</u>, Vol. 55, No.2, February 1985, pp 62-64.

3. Gedye, Rupert, "Works Management and Productivity", William Heinemann Ltd., London, Great Britain, 1979, p 116.

4. Isaak, Merlyn, "Contractor Quality Control: An Evaluation", <u>Journal of the Construction Division</u>, ASCE Vol. 108, No. C04, Dec. 1982, pp 481-484.

5. Kusayanagi, S. and Hately, B.P., "Look Again at Quality Circles", <u>Civil Engineering</u>, Vol. 54, No. 4, April 1984, pp 65-67.

6. O'Brien, James, "Quality Control: A Neglected Factor", Civil Engineering, Vol. 55, No. 2, February 1985, pp 48-49.

7. Ramsey, Tom, "Quality Control 'A Necessity Not an Option'", Journal of Construction Engineering and Management, ASCE Vol. 110, No. 4, Dec. 1984, pp 513-517.

QUALITY IN THE CONSTRUCTED PROJECT:

A WORKSHOP REPORT

by C. B. Tatum [1]

ABSTRACT

In the fall of 1984, ASCE conducted a workshop to determine what actions are necessary to improve quality in the constructed project. The engineering segment of the two-day meeting included sessions focusing on: 1) meeting the owner's ofjectives, 2) achieving teamwork in design, 3) dealing with external factors affecting quality; and 4) assuring quality in the design phase. Sessions in construction portion included: 1) organizing and managing for quality construction, 2) handling conflicts and disputes, 3) dealing with constraints affecting construction quality, and 4) performing quality construction. This paper reports the recommendations for actions by ASCE, owners, engineers, and constructors which resulted from each of these sessions. It also develops implications for quality programs and conclusions.

INTRODUCTION

Improving the quality of the constructed project is an important opportunity for civil engineers. With increased quality come increased productivity and cost effectiveness, along with decreased risk.

Recognizing this opportunity, American Society of Civil Engineers' President, Richard W. Karn, selected quality as a theme for his term. ASCE formed a steering committee, under the co-chairmanship of Holly A. Cornell and Arthur J. Fox, Jr., to plan and conduct a workshop on quality in the constructed product. This paper reports the results of the workshop conducted on November 13-15, 1984. It is based on a summary report prepared by A. J. Fox, session and discussion group reports from the moderators, and the papers presented in the sessions. ASCE will publish a proceedings from the workshop (3).

[1] Associate Professor of Civil Engineering, Stanford University, Stanford, CA 94305

To report on the workshop, the paper first describes the recommended actions. Next it summarizes each of the session and discussion group topics. For the half of the workshop devoted to quality in planning and design these include: 1) meeting the owner's objectives, 2) achieving teamwork in design, 3) dealing with external factors affecting quality, and 4) assuring quality in the design phase. The second half of the workshop focused on quality in construction and included sessions dealing with: 1) organizing and managing for quality construction, 2) handling conflicts and disputes, 3) dealing with constraints affecting construction quality, and 4) performing quality construction. Finally, the paper further elaborates on the discussions from the group on performing quality (which the author moderated) and develops implications for effective quality programs.

WORKSHOP OVERIVEW AND RECOMMENDATIONS

The workshop included 27 papers and written discussions by over 100 participants. Working in discussion groups, the conferees defined recommendations of actions to be taken by ASCE and by owners, designers and constructors toward better achievement of quality in constructed projects.

Eight specific recommended ASCE actions, developed by the steering committee and listed below, were immediately carried forward by President Karn for consideration by appropriate committees of the Society:

1. Prepare policy statements and publish a manual of professional practive for quality in the constructed project, including sections on:

    a. Definitions, guidelines and procedures for relating to owner's objectives and expectations.

    b. Peer oversight project review.

    c. Clarification of project team members' responsibility, authority and liability.

    d. Define and upgrade the authority and responsibility of the resident engineer.

2. Develop guidelines, criteria and standards for automated interfaces for information transfer between participants in the construction process.

3. Develop guidelines, criteria and standards for evaluation of computer software.

4. Reappraise ASCE Manual 45 to base fees on life cycle costs.

5. Quantify measures and benefits of quality oriented planning, design and construction programs and promote learning and disseminating of lessons from performance - successes and failures.

6. Continue and expand the public information program; publish and disseminate a succinct report of this workshop; establish a quality recognition program.

7. Cooperate with other societies and groups to promote quality improvements.

8. Upgrade codes and standards and work through ASCE local sections to get them adopted and administered effectively.

QUALITY IN PLANNING AND DESIGN

This section, and the following portion of the paper on quality in construction summarize the results of eight discussion groups at the workshop.

The report from each group (included in the ASCE proceedings) included recommendations for actions by owners, engineers, contractors, and ASCE to improve quality in the constructed project.

Meeting the Owner's Objectives. - The owner is defined as client, user, contracting officer or any other party or agency who requested or pays for planning and design services. The owner's objective is to obtain high quality facilities, with satisfactory performance over an entire life cycle through good use of planning, good design, good engineering, good construction practices and good management. The architect-engineer's objective is to maintain good relationship with the owner, make a profit, and provide the quality and service from the project that will lead to repeat business.

Good design meets all of the owner's requirements; it is functional and asthetically pleasing; it is cost-effective to acquire, own and operate; it is well coordinated and readily biddable. It also limits errors and omissions to an "acceptable level."

Quality involves: satisfying the owner's worthwhile objectives and meeting expectations (which can be more critical); satisfying the A-E's worthwhile objectives; satisfying the public's worthwhile interests in the completed project. L. S. Garrett, the speaker for this session of the workshop, emphasized several types of owner expectations, many of which are subtle. For example: federal owners must meet the Congress' specific interests and as funding constraints; agencies may have specific objectives regarding facility siting and appearance; design agencies must comply with requirements for distribution of the work and with statutory limitations on design costs.

Satisfying the needs and wants of all parties in a construction project depends on promoting a clear understanding of those needs and wants at the outset and sustaining that understanding through the course of a project by developing and maintaining trust and confidence between the parties. Communication is vital in maintaining proper relationships. It can be verbal at early stages but must then be put in writing. The owner must state both criteria and expectations. Owners often are not sure (or able to visualize) needs. It may be up to the A-E to help the owner state agreements as the quality and estimated cost of project to be designed and built. The A-E, in turn, must describe his plan for the project and prepare quality design, plans and specifications that communicate to the builder precisely what is to be built.

Achieving Teamwork in Design. - The project team includes the designers, contractors, subcontractors, materials and equipment suppliers and sometimes specialty consultants. Coordinating team efforts depends on a project manager who can align the needs of each team member with the objectives of the project. Definition of scope of services and interfaces between team members is an important starting point for this coordination. There should be written delegation statements.

Teamwork also requires a project plan, which defines a strategy for both designing and building. The plan needs schedules, both intermediate and detailed, it needs milestones and checkpoints; it should have "freeze dates" set, after which there will be no design changes made. Teamwork also requires emphasizing designer-constructor communication. This begins best in the planning-design phase with construction input to design. It should continue through construction and startup. During the discussions, the owner was said to be best served if the design team manages construction. There was general accord on the need to have designers see to the faithful execution of their designs by constructors. There is also a strong need for feedback of construction experience to designers by constructors.

The discussion group on design teamwork suggested: use of the quality circle idea in some form; regular and frequent conferences, or conferencing by wire; formal staffing for constructibility review; and formalized feedback to designers from field experience with their designs. It also stressed the need for payment by the owner for quality design and favored prequalification and negotiation over competitive bidding for design services.

Dealing With External Factors Affecting Quality. - Rather than damn the body of codes, standards and regulations, workshop speakers looked at positive as well as negative aspects of such influences on engineering and construction projects.

Looking beyond the client to whom the client in turn serves, it is often the public that must be satisfied with the quality of a project. It is in the public's interest and safety that codes, standards, and regulations are written. Our society seems at times to participate too much, but our increasingly participatory society is a factor in both design and construction. This is not to say engineers or constructors should simply accept regulations as given. They must cooperate in their improvement and refinement.

The working team on this subject of constraints broadened this topic to include (in addition to codes and regulations): fees, indemnification and "image". In this context image refers to the selling of project merits to the owner and the affected public. Good engineering needs to be sold to owners and other involved parties as their best investment to achieve quality in the constructed project. Engineering service should continue from design through full-time construction inspection to startup and initial operation. Recognizing this, fees as set forth in ASCE Manual 45 might better be based on life-cycle project costs rather than construction costs.

Liability claims are seen as the influence on quality in design. Owners should recognize and accept the risks that are not within control of the designer or the contractor.

Assuring Quality in the Design Phase. - Computer-aided engineering is a major asset toward quality assurance - now and increasingly in the future. Computer-aided engineering introduces some new quality problems, but, when properly managed, it can contribute to project quality. Software verification is a major issue. There is need to establish criteria and test methods for computer software. Expanded and improved computer use in design and in the automated exchange of information between project team members presents another important opportunity to assure quality design.

If the principal causes of quality problems include specs that are incomplete or out of date, faulty calculations, incomplete or superceded drawings and design changes that are not transmitted to all parties; all may be improved through computer use. This team also noted a strong need to update codes and standards, and added that ASCE should work on their effective adoption and administration.

Research is needed to provide measures of functionality, constructibility, and maintainability during early design stages. Designers and constructors can better learn from failures and disseminate those lessons; the Architecture and Engineering Performance Information Center at the University of Maryland rates support and cooperation.

QUALITY IN CONSTRUCTION

Organizing and Managing for Quality Construction. - Some more sophisticated owners or contracting agencies in effect purchase design and specifications, constructing the project without significant involvement of the ~~engineer~~ [project mgr.], who is nevertheless expected to retain responsibility for quality of the completed project. Quality will suffer if a resident engineer and his staff are not present and capable of influencing construction-related decisions as well as interpreting the design.

Only the contractor has control over quality of the field work, and field forces can be more motivated by production goals than by quality goals. The strong resident engineer has field experience and is there not to defend the design and specifications but to interpret and modify them according to actual conditions experienced during construction. The importance of the resident engineer suggests that ASCE should prepare position description and qualification statements for different types of projects.

Owners' legal staffs tend to prepare contracts that separate the parties (for easier assignment of responsibility when a mistake is discovered) rather than join them together into an effective team to avoid mistakes. As risks of construction have been passed from designer to contractor to subcontractor, the role of leadership and responsibility for quality have been passed on as well. The owner, or whoever is leading the project team, must assign responsibility, authority, liability, contractual relationships and compensation arrangements. This includes deciding on the desired level of quality and thoroughly communicating the requirements which result from these objectives.

The report from the working team on organization and management stressed the need for flexibility in structuring the project team. However it is structured, the team must provide construction expertise in the very early stages of project planning and conceptual design.

Handling Conflicts and Disputes. - The argument is made that attempts at meeting quality requirements have resulted in increasing claims, increasing litigation, increasing costs and less quality in the constructed project. Independent third-party review of planning, design and construction could possibly do this job better. The working team report on this subject recommends third-party review of all projects over $5 million in cost. This review should start at the time contract drawings are 30% complete.

Pressures of reduced time and budget during the planning and design phase combined with "fast track" construction can lead to litigation during construction. Choice of a contract type (cost-reimbursable, unit-price, lump-sum) inappropriate for the project can be a large factor leading to disputes. Property-type insurance should be developed to cover the entire team and reduce disputes stemming from contract provisions.

Dealing with Constraints Affecting Construction Quality. - Proposition: No constraint, real or perceived, will unavoidably affect construction quality unfavorably. There is an team obligation to produce the quality project despite constraints, which should be viewed as challenges to be overcome. The National

QUALITY IN THE CONSTRUCTED PROJECT

Environmental Policy Act brought challenges, but also improved project quality when broadly viewed. Safety regulations actually enforce good business practices and may save money. Equal opportunity laws may benefit construction in the long run by drawing new people to the industry.

The U. S. Army Corps of Engineers recently completed a study of managing construction quality. One major finding was: "Although we deliver quality projects, quality does not drive the engineering-construction system... Management emphasis is to meet financial goals, award contracts, stay within budget ceilings and schedule milestones. Quality is achieved within the constraints of these factors and not vice versa."

The working team on this subject of dealing with constraints recommended selection of the best contract form for each job; constant communication with the project team; communication to the media and the public to assist in fuller understanding of the project; and active encouragement of innovation.

Performing Quality Construction. - The opening paper in this session (2) emphasized that quality is not inherently goodness, nor excellence, nor expensive. It is first identification and then rejection of work that is nonconforming. Quality criteria must be set -- clearly defined and verifiable -- then conformed with.

The contractor must work in accordance with plans and specifications, perform various quality verifications, inspections and tests, and turn over a quality project. The contractor can perform quality control for the owner if given clear definition of quality requirements, the acceptance criteria and the inspection procedures.

The second paper in this session (1) stated that the traditional approach of having general contractors warrant all work to be in conformance with the contract is failing. Forced isolation of contractor and engineer is also not meeting quality objectives. Although the traditional approach assumes relatively few changes during construction, design-related changes are actually common. A study (1) shows that 20% of major contract disputes in highway work result from changes in design.

Common contract forms have the engineer performing a limited role as the owner's advisor. These contracts may fail because of the importance of active communication and shared responsibility during each phase of the project.

The working team on performing quality construction called on designers to produce explicit specifications, including quality requirements, and to establish the mechanical communication links with construction for common data base use and for information exchange. Contractors should make early and effective inputs to designers; assist in defining quality requirements and methods; assign specific responsibilities for quality to line management and supervision -- to "build in" quality rather than trying to inspect it into a project. Owners must face up to trade offs between cost, schedule and quality. Owners can improve industry performance by establishing quality performance as a basis for contractor selection thereby making it a contractor strategy for competition.

Results of this discussion group, not surprisingly, emphasized the necessity for clear definition of quality requirements and availability of the resources necessary to perform quality construction. With these requirements and resources, contractors can implement programs which assure conformance. Key actions for this implementation include: 1) identifying and preparing necessary work instructions or procedures, 2) communicating these instructions to construction supervision, and 3) planning and conducting the inspection and testing activities necessary to assure compliance with the requirements.

IMPLICATIONS FOR CONSTRUCTION QUALITY PROGRAMS

The recommendations from this workshop result in several implications for construction quality programs. To achieve quality in the constructed project, quality control programs should:

1) Incorporate all stated owner objectives and understood owner expectations. The latter frequently stem from other organizations influencing the owner, such as regulatory agencies, local governments, sources of financing, and local suppliers and contractors.

2) Begin in the early planning or conceptual design phase of a project. This will allow construction input to quality requirements. Early construction thinking regarding both how the project will be built and how the quality requirements will be met can assist in defining realistic requirements. Construction quality personnel can provide the following types of beneficial input:
   A. Which types of quality methods are preferred.

   B. What acceptance criteria are realistic.

   C. How much documentation is possible.

   D. What changes are necessary to incease the inspectability of the design.

   E. What quality assurance requirements can realistically be specified for various types of material suppliers and subcontractors.

   F. How to increase the flexibility given to inspectors in the field regarding acceptance of equivalent, but not precisely conforming work.

QUALITY IN THE CONSTRUCTED PROJECT   13

3) Focus on conformance to requirements by identification and rejection of non-conforming work.

4) Include adequate flexibility to operate under the changed conditions which occur in the field.

5) Obtain owner concurrence that the design and the specified quality requirements meet all project objectives and expectations.

6) Limit inspection interference with construction operations to the minimum necessary to assure conformance to requirements.

7) Evaluate the cost effectiveness of requirements in excess of industry standards.

8) Encourage and facilitate continuing communication to make requirements clear and quickly resolve quality related disputes between construction and engineering, procurement, or startup.

9) Where project size justifies machine-readable data bases for engineering and procurement, take advantage of the information available in these data bases to specify and communicate quality requirements and to resolve quality problems.

10) Include the resident engineer function for interpretation of the plans and specifications at the construction site.

11) Demonstrate the cost effectiveness of quality requirements as a means to gain greater management commitment.

CONCLUSIONS

Discussion groups at the ASCE Workshop on Improving Quality in the Constructed Project concluded that several actions by ASCE, owners, engineers, and constructors will improve quality. The recommendations ranged from broad actions such as preparing an ASCE manual on means to improve construction quality to specific steps for building in, rather than inspecting in, quality. The actions included all phases of projects and each of the project team members.

Three themes ran through the recommendations. First, quality improvement depends on increased definition and awareness. The groups judged full specification and communication of requirements, expectations, methods, and acceptance criteria as an important prerequisite to quality in the construction project.

Second, improved quality will increasingly rely on advanced technology. Recognizing this requires both establishing new types of controls and using the new information and other resources. For example, quality control of software for structural analysis and design is a major new challenge. Computer-aided design systems hold great promise for providing additional information to construction. In addition, these data bases may eventually support the automation of many construction operations to improve quality and decrease cost.

Finally, each of the eight discussion groups advocated a broader view of quality in the constructed project. Each project phase is important. An integrated perspective which recognizes key actions by each team member is necessary to effectively meet quality objectives.

REFERENCES

1. Hester, Westen T., "The Traditional Approach to Assuring Construction Quality," presented at the ASCE Workshop on Improving Quality in the Constructed Project, Chicago, November 13-15, 1984.

2. Kulchak, R. E., "Performing Quality Construction," presented at the ASCE Workshop on Improving Quality in the Constructed Project, Chicago, November 13-15, 1984.

3. "Proceedings of the Workshop on Quality in the Constructed Project, November 13 - 15, 1984, Chicago, ASCE, in press.

ACKNOWLEDGMENT

The participants in this workshop make important contributions to the discussions and the recommendations of the working groups. The steering committee, identified in the workshop proceedings, provided leadership for the sessions. A. J. Fox Jr. summarized reports from each session into the executive summary which provided a basis for a portion of this paper.

Quality Management in the US Army Corps of Engineers: An Evaluation

Homer Johnstone, PE, Brig. Gen. US Army (Ret.)*
ASCE Member

Abstract

Quality management programs are employed to assure the achievement of specific standards of performance. In the construction industry, quality management personnel face common problems in their efforts to deliver quality products, and successful quality management programs share several key basic principles. The US Army Corps of Engineers recently evaluated its quality management program to identify its strengths and weaknesses. A discussion of this analysis follows, including recommendations for improvement which may be applicable to other quality management programs.

Introduction

Quality is not the automatic end-product of good designers and constructors working together. Nor is it necessarily only the result of divine intervention. Depending upon the personalities involved, materials at hand, and nature of the project, high quality construction may be delivered on a hit or miss basis. We are learning, in the industry, that if a specific level of quality is desired, there must be a formal mechanism operating to assure the achievement of that level of quality. We can call this mechanism quality management.

How well do these quality management programs work? That important question can be answered by taking a hard look at one. A thorough analysis of an existing quality program can identify its strengths and weaknesses, and opportunities to improve it. Those developing or revising quality programs can learn from the experiences of others.

Overview

The US Army Corps of Engineers decided a number of months ago to evaluate its quality management program. The Chief of Engineers shared concerns expressed by senior commanders about the quality of some of the Corps' work. He commissioned a Blue Ribbon Panel to evaluate the effectiveness of the Corps' quality assurance/quality control program and make specific recommendations for improvement. The Panel was

*Principal, Quality Management Associates; and Chief Operating Officer, Bentley Engineering Company, 560 Mission Street, San Francisco, CA 94105.

15

directed to find methods to improve the quality assurance/quality control operations, the interaction of the design and construction functions, and field administration procedures in general. As a benchmark for the study, the Panel defined managing quality construction as the ability to construct projects according to professional quality plans and specifications in a timely fashion within budget.

The study focused on the Corps working units in charge of construction management. These are the resident or area offices composed of field personnel, headed up by a resident or area engineer. The field personnel in these offices are directly responsible for actual construction of Corps projects, while engineering design is managed by separate, district offices. Workshops, interviews and surveys were conducted to query more than 60% of the Corps' field personnel. In addition, actual job sites were visited to review Corps work. The study also focused on those who use the product. Nearly 200 of the Corps' customers or users were surveyed about the quality of Corps projects.

The overall finding of the Panel was that, in general, the Corps' quality assurance/quality control program is working. It is founded solidly on appropriate and effective principles of quality management. It reinforces the concept that producing a quality product is the contractor's responsibility, but does not relieve the Corps' inspection force from the absolute responsibility of assuring that a quality product is delivered. It encourages contractors to become familiar with the Corps' standards and technical provisions. And, it augments the resources available to the Corps by requiring the contractor to perform specific tests, and certain other surveillance and inspection functions, where well-defined measurements are obtainable. Conversely, the panel concluded that the program does not substitute for adequate Corps' quality assurance inspection, particularly where parameters are not readily verifiable. Further, it does not diminish the resident or area engineer's responsibility to apply early and vigorously the full range of contractual remedies available to him when quality is lacking.

The weaknesses the Panel identified related mostly to the implementation of the program. The soundness of the concept was rendered useless in some cases because personnel did not follow it, were not induced to follow it, or were actually induced to <u>not</u> follow it.

The Corps' quality management program, while perhaps difficult to implement, is relatively easy to describe. Contractors of Corps projects are required to develop and follow a Contractor Quality Control plan, called the CQC, involving a formal, three-phase inspection process. It is the responsibility of the Corps resident or area engineer in charge of construction to independently monitor, test and check the CQC programs -- this is the quality assurance, or QA, facet of the program. Several methods of dealing with unsatisfactory work are available to quality managers such as requiring tear out of deficient work, witholding payment, issuing a formal, unsatisfactory contractor rating, etc.

CORPS OF ENGINEERS

Discussion of Major Findings

The Panel found that the majority of the Corps' customers, users, and field engineers were generally satisifed with Corps construction quality; however extensive problems were identified with some aspects of the program and at some specific sites. The Panel concluded that the overall QA/CQC function could be significantly improved at all levels throughout the Corps. Although quality products were delivered in most cases, the Panel found that quality did not always drive the engineering/ construction system. Management's emphasis on financial goals placed constraints on the achievement of quality. Field personnel were rewarded for meeting time, cost, and budget criteria. Quality was achieved within these limits, and not vice-versa.

The inconsistent and sometimes negligible emphasis awarded quality by Corps' management sent a message to the field that quality assurance was a low priority. Only a few Corps units had a formal quality assurance element. Quality assurance rarely appeared in the area or resident engineer's job description, although he is directly responsible for assuring quality construction.

A lack of understanding of their critical quality assurance responsibilities pervaded the field personnel, and there was little shared perception about how a quality assurance program is supposed to function. Policies and degree of implementation varied from office to office, with the duties, responsibilities, and skills of quality assurance personnel differing greatly. Contractor quality control was more formal than quality assurance. Quality assurance was ad hoc, built upon informal relationships between Corps representatives and contractors. There was no guide to the degree of quality assurance necessary for different CQC plans. CQC plans were frequently shelved. Contractors wrote and submitted the plans for contract compliance, not as a field tool to check quality during construction. Corps quality assurance personnel were often unable to react effectively, or chose not to react, to unsatisfactory contractor performance.

Some quality assurance personnel lacked requisite technical ability and contract interpretation skills. Others felt that support was lacking from the Corps' higher authority. They believed that contract administration worked against them, that headquarters usually supported the contractors on contested quality claims, and that the evaluation system encouraged minimal claims and costly resolution of problems rather than contract performance.

Manpower problems were cited, relating particularly to training of quality assurance personnel. Key quality assurance personnel would not take time from the job to travel to the centralized, QA training program.

Additional personnel-related problems diminished the effectiveness of the Corps' quality management. There was an unsatisfactory level of friction between engineering and construction units. Field personnel felt that the engineering staff was basically unresponsive to their concerns, and assigned low priority and insufficient funds to design support during actual construction.

There were few formal opportunities for field personnel to contribute to the design process. Often, their comments were ignored. Other field personnel were unwilling to seek or heed the technical advice of designers.

The Panel identified at least two problem areas affecting quality which may be more prevalent in the Corps than in private industry. The fourth quarter push to award contracts reduced time for adequate constructibility reviews, and further strained relations between engineers and field personnel. Additionally, an excessive amount of administrative paperwork was impeding area and resident engineers from performing critical quality assurance responsibilities.

Surveys revealed some shortcomings in the Corps' responsiveness to users and customers. The major problems areas users identifed were: completing a facility on schedule; assuring the operability and maintainability of the facility; providing comprehensive operation and maintenance documentation; providing training for operations and maintenance personnel; and providing timely correction of deficiencies.

Discussion of Recommendations

The Panel identified the strengths of the Corps' quality management program and many opportunities for improving the program. In a broader sense, its recommendations highlight the features integral to any effective quality control program in the construction industry.

Most of the problems related to ineffective implementation of the existing program. The Panel felt it was important to educate Corps' quality assurance personnel about the strengths and weaknesses of the program -- to share with them the rationale and impetus for making it work more effectively. The most important recommendation the Panel made was that Corps management visibly renew the emphasis placed on quality assurance. In the same way ambivalent policies adversely affect field personnel, a visible commitment to quality management encourages and supports the efforts of field personnel. The Corps' management must motivate Corps' employees to implement quality assurance, and Corps' employees must motivate contractors to follow CQC plans. This renewed emphasis gives field personnel the confidence to act effectively when contractor performance is unsatisfactory.

Thorough CQC and QA inspection reduces costs both for the Corps and contractors. Quality assurance personnel must convince contractors that effective CQC is beneficial to them, and that deficient construction will eventually be detected by a QA inspector. Quality assurance personnel must communicate with the contractor before the fact about what is expected.

The Panel recomended that the Corps improve programs to acknowledge good contractor performance, and adopt a philosophy and support system which would allow quality assurance personnel to effectively confront unsatisfactory contractor performance, and utilize poor contractor ratings.

The Panel suggested several ways to enhance the qualifications and abilities of quality assurance personnel. They must be given adequate training in their areas of expertise, for quality assurance personnel must be more versatile today, proficient in several areas of the quality assurance process. Additionally, training in the Corps' QA/CQC process which does not require extensive time away from the job must be available, such as video packages, and area and resident engineers should be encouraged to take advantage of this training.

Several recommendations focused on reducing to an acceptable level the administrative load of key quality assurance personnel. Although not as extensive a problem in private industry, the message behind the recommendation is relevant: The person in charge of quality assurance must have available the time and resources necessary to effectively carry out his responsibilities. Management is merely giving lip service to quality assurance if chief QA personnel are required to perform a multitude of support functions at the expense of quality. Unnecessary emphasis on areas peripheral to quality should be reduced. The Panel recommended that the Corps make available to field personnel more state-of-the-art technology to facilitate project administration, such as computerized scheduling and budgeting, and encourage the judicious use of administrative support consultants.

The nature of the engineering-construction interface can greatly affect project quality. Field personnel must be given formal opportunities to comment on the constructibility of the design at several stages. A project should not be advertised until the design has been reviewed. Design engineers must be more accessible to field personnel during actual construction. The Panel recommended a more formal system in which constructibility comments are accommodated before the project is advertised. Designers should meet formally with field staff to review design intent, assumed foundation conditions, contract operation, key quality assurance items, and shop drawing procedures. Good, informal relations between engineering and construction units should continue during the construction phase.

Engineering must allocate more resources for the post-construction phase, as well as the construction phase. This is when users' impressions of the Corps are most significantly affected. User satisfaction is very important; but it should be viewed as a goal in tandem with effective quality management, not a substitute. The Panel recommended that the Corps improve techniques to solicit, use, and document user input throughout project development. Engineering should be available to verify that all systems are operating properly. They should be present at pre-final and final inspections to accommodate user complaints. Accurate operations and maintenance manuals are critical, and should be included as a pay item in contracts whenever possible.

The Panel also recommended that the Corps improve project documentation so that experience, innovations, and organizational successes and failures could be shared by others in the Corps.

Summary and Conclusions

In summary, the Panel found the basis of the existing Corps quality management program sound, but recommended, among other things:

o   A renewed, visible commitment to quality assurance from the Corps' top management.

o   "Back to basics" revision of the QA/CQC program to be shared with all QA personnel.

o   A tough corporate policy demanding quality from a less-than-willing construction system.

o   A new, comprehensive, decentralized training program for QA personnel.

o   Reduction in administrative paperwork requirements for key QA personnel.

o   Formal opportunities for field personnel to comment on design constructibility, and increased engineering availability to field during construction.

o   Increased engineering respsonsiveness and availability during post-construction and initial occupancy phases.

o   Improving techniques to solicit and accommodate user input throughout project development.

The ongoing implementation of these recommendations is significantly altering the way the Corps does business. The quality management function will be more effective, and therefore the quality of Corps construction projects will improve. The Corps and its QA/QC system may be unique in many ways. Most basic principles in the Corps quality management program, however, can be shared by private industry programs. The problems the Panel identified are common, and the solutions, as well, are applicable to quality management programs everywhere.

of a project constructed at a rate of a typical floor every three days without a day of delay related to quality. Instead of being wasteful in materials to assure concrete strength, we applied the latest in technology and saved about $4.00 per cubic yard on nearly 100,000 yards of concrete. Along the way, we set a world record for single shot pumping concrete 1,007 feet vertically.

There are a number of things which must happen at the start of the project. The first and foremost is that project management must recognize that QA is not another chain in the system to obstruct the construction progress. This commitment must be shown not only by word, but also and most importantly, by actions in support of the program. Quality management can save money. We don't do anything cheaper the second time than it can be done the first time.

The first action is to become intimately familiar with the project. This includes the contract documents and the planned construction program. The two must be made to fit. Painstakingly the drawings and specifications must be reviewed to locate the problem areas. These must be located and alternative solutions developed and carefully presented to the design team for their approval. When the design team recognizes that the contractor is making a conscientious effort to build the job properly, they will be cooperative. We were fortunate to have such support. Two major problems were prevented on our job through such cooperation:

- Our 9'9" thick mat was to contain 16,000 cubic yards of 6,000 psi concrete with casting to be done in two pours. Five layers of #11, 14 and 18 bars were on the top and bottom. The steel density at the top was so great that we could not use the needed 2 5/8" vibrators. Tremies were not even considered. A full scale drawing of a section of each layer was made on tracing paper. These drawings were placed over a light box and each one moved in a sequence to see if one combination would open space. After many hours of careful study, it was found that by moving the top layer of #14 bars 1/2 space, windows allowing clear access were opened. CBM Engineers approved the change and we used 12 pumps to place 8,250 yards of 6,000 psi concrete at a rate of 850 yards per hour.

THE ROLE OF THE QUALITY ASSURANCE MANAGER
ON A LARGE COMMERCIAL PROJECT

Paul Little
Turner Construction Co. of Texas
Houston, Texas

The role of the quality assurance manager on a high rise project being built on a tight time schedule is a challenging and diversified assignment. It is made especially difficult when there is no history for such an assignment. The job description is simple: See that it's built right and save money. The daily tasks range from being the local authority on the use of everything from materials to testing procedures. Besides the technical aspects of the assignment, the QA manager is faced with providing psychological counseling for a frustrated lab technician, a concrete batch plant operator or an iron worker. He must be all things to all people while standing up for quality. He must be somewhat like being an Army Chaplin.

My first experience as a QA manager came under severe fire. I was selected to meet the challenge for the construction of the sixth tallest building in the world - the Texas Commerce Tower in Houston. From a construction viewpoint, plans called for a composite tube taller than any built - and the tower would rest on a record size foundation.

This program started before I arrived on site. The project management had noted potential problems with the large amount of 6,000 and 7,500 psi concrete. Such strengths were not common in Houston at that time and project delays were being encountered with lower strength concretes.

During early project discussions, the recommendation was made that a Turner construction engineer be trained as the on-site quality manager. Efforts were to be directed to problem prevention rather than face later delays and costly redo. The cost to the owner in interim finance and to the contractor in overhead for each day of delay was astronomical. A principle factor for not going to an outside organization for the service was that the QA manager had to be a part of the organization responsible for the results. He had to have authority and direct access to workmen and to suppliers and subcontractors.

The result of this decision was successful completion

QUALITY ASSURANCE MANAGER 23

- The second case involved the structural steel spandrel beam and the 7,500 psi concrete. Engineers seldom leave much room outside of the steel to place concrete. With our program of pumping the concrete through the core and obvious safety consideration, it was necessary to place and compact the concrete from the inside of the building. The spandrel beam was so close to the inside form face that we could not properly place the concrete and use the appropriate vibrator. A small vibrator could not do the job. We studied the problem with the engineers. They found that by changing a few details, the beam steel could be moved closer to the outside face and we no longer had a problem.

These findings point to the need for consideration of both design and construction criteria. Each is important. Construction related problems must be uncovered before the preparation of shop drawings. Once the job is underway, it is too late to economically make changes. Locating problems is a management responsibility. Solutions should not be left to the workers casting the concrete. Who, but the QA manager, is going to take the time on a fast moving job to consider what might appear as small details months before the work will be done?

Once the "snakes" are out of the job, the QA manager must prepare a plan. A flow chart of the processes involved is helpful in locating where controls must be inserted. You have been provided a chart.

Suppliers and subcontractors must be brought in on some of the planning. Some will see its value while others will look upon it as eye wash. This latter attitude must be changed. Friendly persuasion often produces the desired results. If not, explanation of the economic consequences of delaying the job one day often gets attention. In hindsight, setting up a QA program is best initiated before purchase orders are issued. Then it becomes a contractual requirement. We did it the hard way.

The plan must be in writing. It doesn't need to be complex. We used a series of short statements and a checklist of responsibilities. When we were pumping lightweight concrete 1,007' straight up, we had to carefully orchestrate over 100 people working for several suppliers, subcontractors and laboratories. Each person knew that he was part of the quality team and that others were depending upon him.

Once the job is underway, the QA manager is busy with follow-up. He must verify that every party to the project is doing his job. Intense follow-up is necessary in the beginning to get everyone use to the process. If a weakness in the system develops, the cause of the problem must be removed or a special quality check placed at that point. The QA manager cannot assume anything. Many organizations and people in the construction business do not understand the technology of what they are doing. These must be located and educated or replaced.

Communications must be established at all levels. The QA manager must have his chain of contacts within all related firms. These contacts do not always have to be the senior managers. He must gain the confidence of his contacts and the workers on the production line. He must be ready to pass out compliments. The compliment can often be a greater motivator than a reprimand. Construction workers, like everyone else, like recognition. It is a basic motivator. When a worker does something wrong the first time, question whether or not he was given the proper instructions. Suggestions for improvement given in a quiet understanding way are effective.

QA works. Quality is not always the result of assumptions that you "have a good team of experienced people". Not all experience is good for every job. Every design is different. The production people on the job generally use the long proven techniques. The QA manager must look ahead for better ways to do the job. This has become especially important in this era when time and productivity are the keys to profitable construction. We must build with better technology. Sheer brawn doesn't always work.

A good QA manager should have an adequate technical background. Practical knowledge of physics, construction methods and properties of materials are indeed necessary. However, these must take a back seat to the managerial and communication skills required to motivate a multipicity of characters and personalities to perform as a team.

The QA manager must know what to expect from his materials, methods and people. He should know that his materials vary in consistency of manufacture, and that what might appear as simple pumps and self-lifting form systems can be frustrating wonders of high technology. Most importantly, he must be sensitive to the people that work with these elements. Dedication to the desired result should be a mutual goal of all of the team members. It is the QA manager's job to make the team aware of the desired results and the game plan necessary to achieve them. The implementation of simple systems of networks and procedures must be constantly monitored and updated to insure that all elements of the overall plan are in tact and operating appropriately. Flexibility is required as well. He must have alternate methods of achieving the same results.

The QA manager must have a positive and confident attitude and a personal presence that makes the team members feel compelled to work with him. Finally, the QA manager must not be afraid to expect the best.

CONSTRUCTION MANAGEMENT: DALLAS NORTH TOLLWAY PROJECT

Stephen B. Quinn*, M. ASCE

ABSTRACT: The Dallas North Tollway Extension is a ten mile project consisting of five major construction contracts designed by five consulting engineering firms. The General Consulting Engineer for both design and construction is Howard, Needles, Tammen, and Bergendoff (HNTB). The construction administration staff of HNTB for this project consists of a Project Manager and two Resident Engineers, all licensed in the state of Texas, along with three Assistant Resident Engineers and an adequate staff to properly inspect the work. Working under the Construction Management Team is an on-site independent testing laboratory reporting directly to the Project Manager. The client for the project is the Texas Turnpike Authority (TTA). They provide full engineering liaison and are involved in all major decisions affecting the ultimate cost of the project. The author discusses the overall construction management organization; the liaison process to assure continuity and consistency between the various contracts; the documentation procedure; the overall methods of handling the construction inspection of the project. Also included are the means and methods developed for project liaison with the client. The author further discusses the revised method for inspecting a project of this nature that was implemented and explores the benefits of both the system initially used and the alternate system developed for the inspection of the project.

INTRODUCTION

This article will attempt to illustrate the importance of flexibility of the design and construction inspection process to insure the project is constructed within the time and budget allotted, while at the same time insuring a quality product for the client. The 10-mile Dallas North Tollway is a $168 million dollar bond project. It is divided into 5 major construction contracts, and in addition, a number of ancilliary contracts such as toll facilities, fencing and signing and lighting. The organization to inspect this type of facility and the flexibility necessary to respond to the contractors needs is discussed, as is the format for project record keeping.

---

*Associate, HOWARD NEEDLES TAMMEN & BERGENDOFF, 14114 Dallas Parkway, Dallas, Texas 75240.

CONSTRUCTION MANAGEMENT

PROJECT BACKGROUND

In 1965, the TTA began construction of the Dallas North Tollway. The project was constructed along an abandoned right-of-way of the St. Louis Southwestern Railroad as a limited access urban highway facility. This 9.8 mile Tollway was opened to traffic in 1968. Its location in the Dallas-Fort Worth Metroplex is shown in Figure 1.

The facility provides a north-south connection between I-35 on the south and I-635 (a circumferential freeway) on the north. Although it was desirous to construct the Tollway as a six lane facility for its entire length, the anticipated toll revenues would justify six lanes for the southerly eight miles only, thus the northerly 1.8 miles were constucted as a four lane facility. However, in anticipation of future growth in the northerly portion of the corridor, the TTA, and its General Consultant, HNTB, planned the northern portion of the facility for a future widening to six lanes.

Rapid development in the late 1970's in the corridor, especially in the area immediately north of I-635, generated interest in extending the Tollway northward to SH 121. Again, due to cost and revenue restraints, it became necessary to widen and extend the Tollway in two phases. Phase One, shown in Figure 2, resulted in a $168,000,000 bond issue sold on August 1, 1982. It consists of widening the northerly two miles of the existing Tollway from four lanes to six lanes, constructing an Interchange between the Tollway and I-635 and extending the Tollway along the alignment of the existing Dallas Parkway. The section between I-635 and Keller Springs Road will consist of a six lane toll facility with three lane service roads in each direction. At Keller Springs Road the Tollway will be reduced to four lanes northerly to the Dallas/Collin County line. From this point northward only the service roads will be constructed. This is the project currently under construction.

The total construction cost was projected to be $82.5 million. Engineering, legal services, material testing, and right-of-way acquisition totaled $41.5 million, for a total project cost of approximately $124 million.

DESIGN COORDINATION

In an effort to optimize both the design and construction efforts, the project was divided into five sections.

HNTB was selected as General Consultant (GEC) to the TTA with the responsibility for design coordination of these sections. The number and limits of the sections were based upon many factors, including expediency of engineering design, maintenance of traffic, scheduled acquisition of right-of-way, coordination of utility relocations, and size of construction contracts. The design was reviewed for both conformance to codes and coordination of facilities within the contract limits.

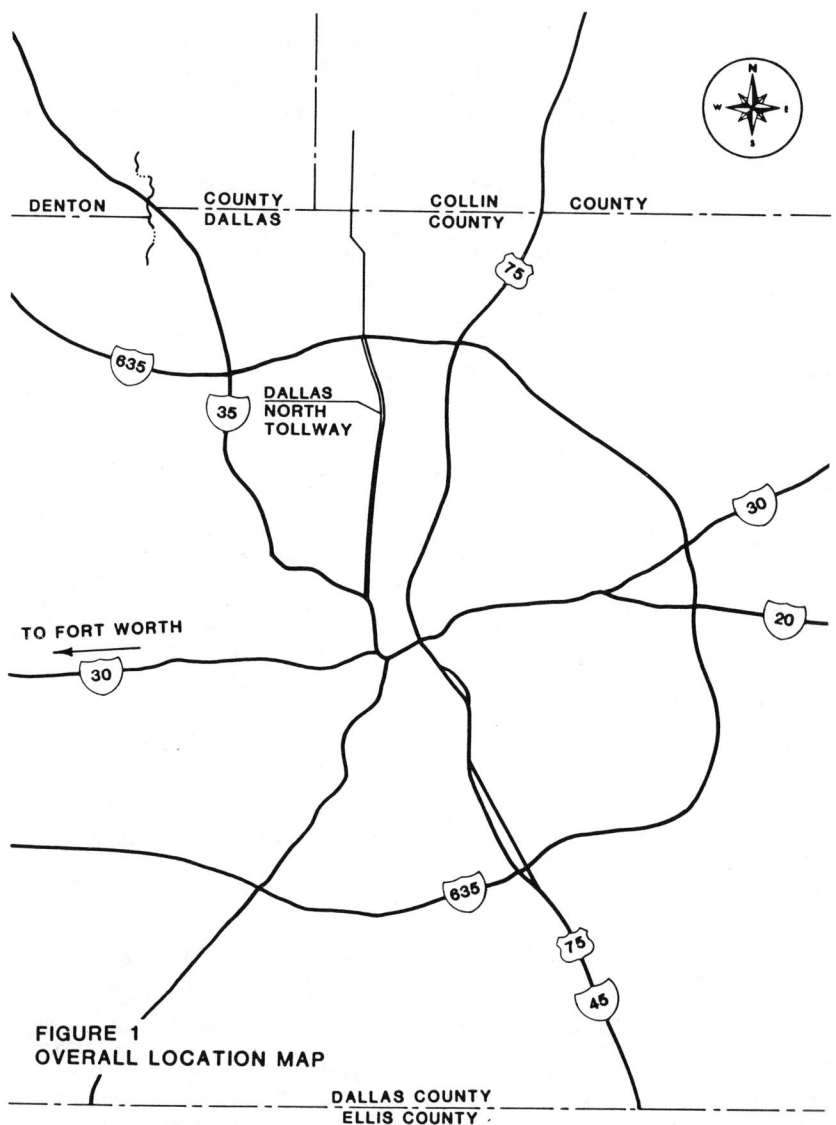

**FIGURE 1
OVERALL LOCATION MAP**

CONSTRUCTION MANAGEMENT

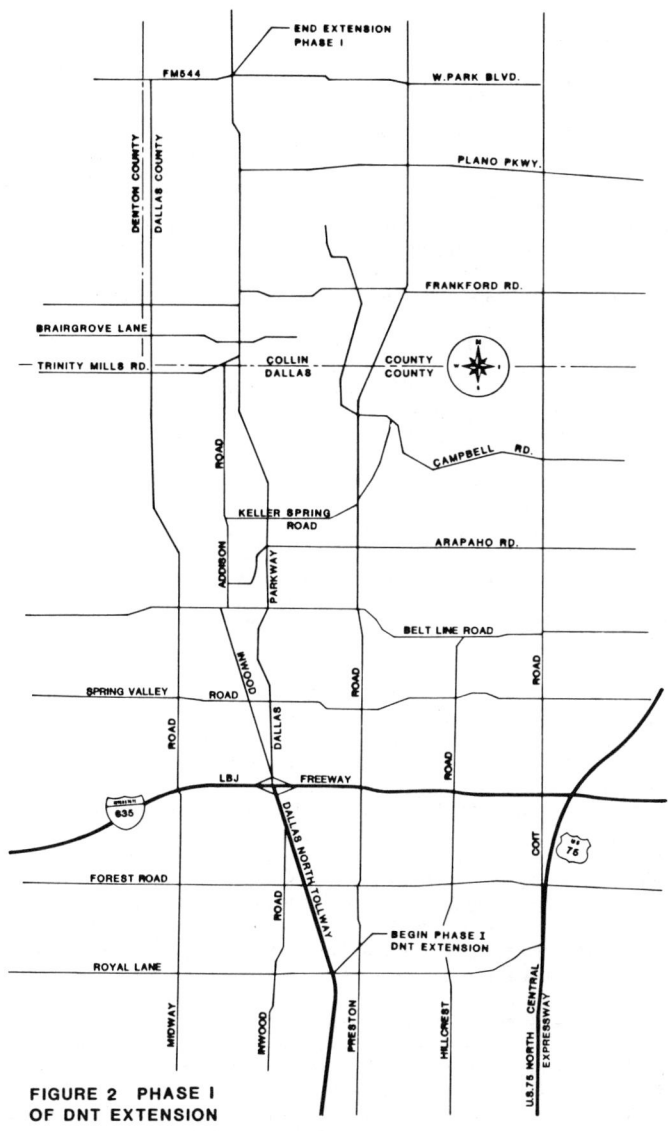

FIGURE 2 PHASE I
OF DNT EXTENSION

The Texas State Department of Highways and Public Transportation Standard Specifications were utilized adding supplemental specifications for job specific requirements to implement the construction of the project.

CONSTRUCTION MANAGEMENT

In addition to serving as GEC for design of the project, HNTB was also selected as Construction Manager providing construction inspection services for the entire project. HNTB provides the Construction Project Manager and licensed Resident Engineers to oversee the construction inspection of the work. The TTA provides a small staff to aid the coordination of the construction inspection and administration. The initial organizational chart, illustrating lines of communication and responsibiility is shown in Figure 3.

One of the first charges of HNTB was to aid the TTA in selecting an independent testing laboratory. The selection committee was composed of two members of the GEC staff and three members of the TTA. After interviewing ten laboratories and short listing five, this difficult selection was made from a number of qualified laboratories in the City of Dallas. This testing laboratory was then put under the direct charge of the Construction Project Manager to aid in performing the materials testing necessary to implement the required standards. Maxim Engineers, Dallas, Texas, was hired to perform the testing on the project. The testing laboratory's function is to provide on site facilities to assure test results are immediately obtainable. The laboratory is responsible for all testing, including gradation of materials, slump tests, concrete cylinders and beams, compaction, etc.

Preliminary Efforts--Initially, HNTB revised an existing construction manual previously written for the TTA on the Houston Ship Channel Bridge to address specific aspects of the Dallas North Tollway Project (HNTB also served as GEC on the Ship Channel project.) This manual delineates the authority and duties of the inspection personnel, equates this with the TTA standard forms, and delineates the requirements for the uniformity of inspection techniques, so that contractors on the different sections would receive consistent treatment from the inspection personnel. The successful construction bidders were given copies of this manual to acquaint them with the reporting requirements of the TTA and the methods utilized to make interim and final payments to the contractor. The TTA's procedure allows payment to the contractor within fifteen days after the end of the payment period.

Upon the TTA's approval of the organization chart (see Figure 3), HNTB embarked on an intensive training program for the staff to assure that the staff had adequate knowledge of inspection techniques and procedures. Aside from the normal daily training that takes place under any inspection program, experts were brought in from both within and outside of HNTB to conduct formal training programs for the staff. These programs included training in the installation of drilled shafts, mechanics of materials used in construction, and installation

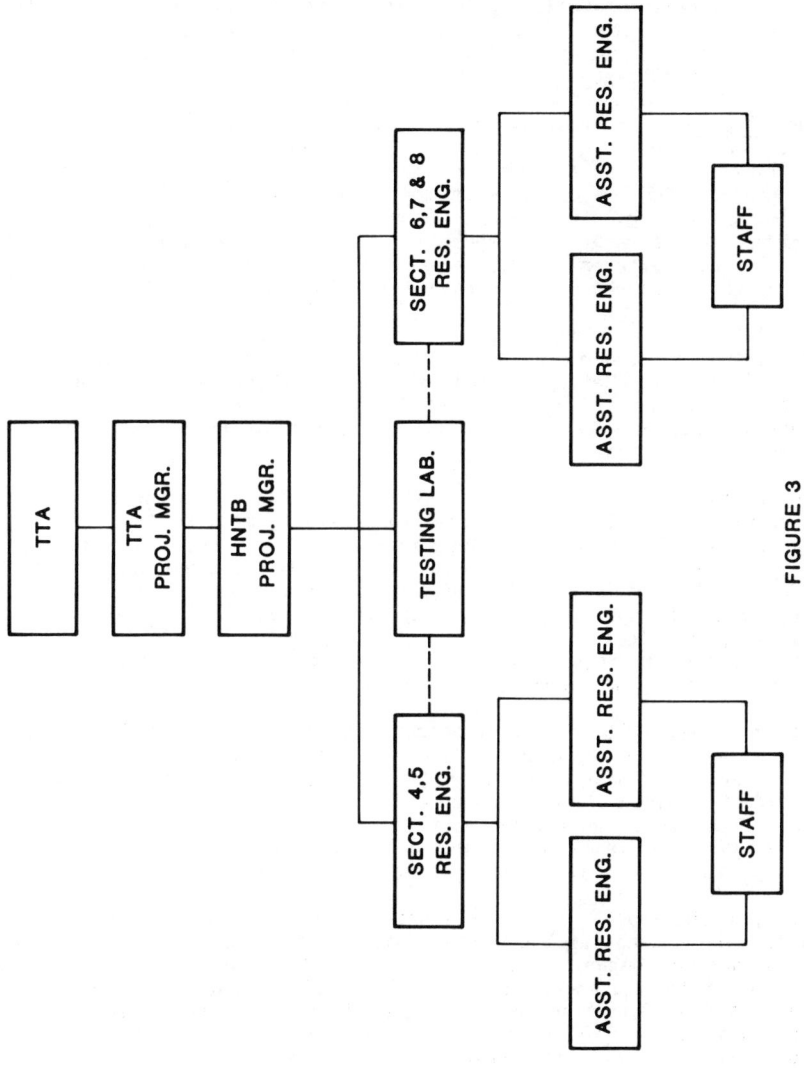

FIGURE 3

of post-tensioning tendons. In addition, the inspection staff visited various manufacturing plants to acquaint themselves with the techniques utilized to manufacture such items as concrete pipe and precast box culverts.

Construction--Construction of the project began in the fall of 1983 with the award of the first contract (widening of the existing Tollway) to Kasler Corporation. Kasler Corporation is a California contractor who has recently become a major factor in bidding in the Dallas-Fort Worth Metroplex. Kasler Corporation was also low bidder on the second section, comprising the major Interchange with I-635 awarded early winter of 1983; and the third section from Spring Valley to north of Arapaho Road, awarded early 1984.

Initially, Kasler Corporation responded to the construction work by providing independent supervision for each of their contracts. HNTB's organization was set up to administer separate contracts with separate Resident Engineers and Assistant Engineers. The inspection system worked exceptionally well. Also, it is relatively the standard for this type of work.

Approximately one year into construction, Kasler Corporation, with approval of the TTA, reorganized their staff, merging the first two contracts under a single Project Manager. As construction proceeded, they determined they they could run a more efficient operation by merging all three of their contracts under common leadership. With over $50 million worth of construction under their direct juris- diction, they felt that utilizing the same supervisors for subsurface, paving, and structural work presented a far more efficient operation then providing individual crews for each of their sections.

In response, HNTB also reorganized its staff under similar product lines. Figure 4 shows HNTB's revised organization. HNTB's Resident Engineering staff was well suited to this reorganization. One of the Resident Engineers' background was primarily in the design, struc- tural, construction and inspection. The other Resident Engineer had extensive experience in paving and underground construction. The reorganization was implemented in November of 1984.

Also, in November, 1984, our fourth major construction contract was bid. The successful bidder for this $15 million contract was Allan Construction Company of Dallas, Texas. This contract was given a Notice to Proceed in January of 1985. HNTB is continually evaluating the staff organization to insure that it is a viable system that can be utilized for the entire project, taking into account we now have more than one contractor present.

In addition to the general work of building highways and bridges, tollways are unique in that they always have a building contractor involved, as well as Toll Plazas, Auxiliary Toll Booths and buildings. To monitor this work, it is presently our intention to place an architectural engineer in responsible charge of the construction inspection of this phase of the work.

CONSTRUCTION MANAGEMENT

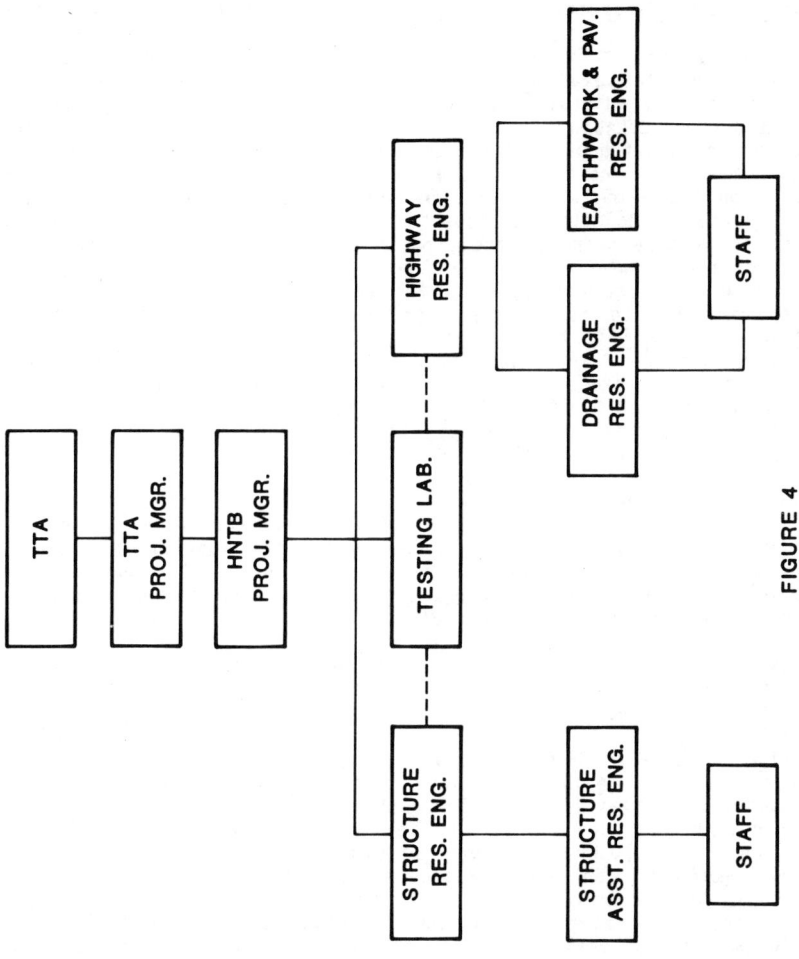

FIGURE 4

Construction Inspection--The personnel assigned to construction inspection for the project have varied backgrounds in the construction technology field. Three graduate Civil Engineers make up part of the staff. One has a background in Highway Design, one in Structural Design, and one in Drainage. Two inspectors have four year degrees in Construction Technology, and one has a degree in Geology. The remainder of the staff have hands-on experience in construction inspection. Assignment of personnel to the various construction operations is based on their individual area of expertise. In addition, they are given exposure to all other aspects of the project to gain familiarity with the entire scope of the project before being assigned to areas outside their own field. As a general rule, each work crew or group of work crews is assigned an inspector. It is the inspector's responsibility to assure the work being performed conforms to the plans and specifications. The inspectors' work is periodically reviewed by their immediate supervisor. To assure fairness to the contractor, questions are initially addressed by the field inspector through the chain of commmand to the project manager. If the contractor is still dissatisfied with decisions, a meeting is set up with TTA officials to review the problem area.

Frequent meetings are held with the contractor's personnel. The subject of these meetings may be job progress (held weekly), individual job problems or discussions relating to upcoming operations. Personnel attending these meetings include the contractor's senior staff and the inspection personnel assigned to the operation, supplemented by HNTB's senior staff.

Staff meetings are held periodically to discuss various inspection techniques, job assignments, technical aspects of the project, vacation schedules, etc.

Design Liaison--As GEC for design, HNTB is serving as design liaison to the Section Engineers during construction operations. HNTB is responsible for aiding the construction management team in assuring the client that the project is being constructed according to the design intent. When design questions arise in the field, the Construction Resident approaches either the GEC Design Highway Project Manager or the Design Structural Project Manager. The Project Managers then decide whether the solution requires input from the original section designer or can be handled by the GEC staff. This system has been in use for well over a year now, and has generated respect for both parties. The field personnel have a good line of communication with the design engineer, and the design personnel are assured that the field personnel are capable of attaining the design intent during the construction process.

Documentation--The major charge of any field organization, in terms of construction management or construction inspection, is to produce the accurate job records necessary to verify that the contractor has conformed with the plans and specifications. HNTB has implemented the following procedures to assure this compliance:

CONSTRUCTION MANAGEMENT   35

1. Diaries

Each Inspector is required to keep a daily diary. It delineates the work inspected that day. It also includes weather conditions, the name of the contractor and subcontractor, the size of their crews, and any remarks he may wish to incorporate into the project record.

2. Daily Reports

Daily reports are a compilation of all diaries of the inspection personnel. They are an accurate history of the work that has taken place on any particular day. They include a detailed record of the contractor's crew size, his equipment, any subcontractors working on the project, the work performed that day, including locations of any finished installation, and finally the payment quantities placed during that period. These reports also include any major item of note. They are signed by the Resident Engineer and initialed by the Construction Project Manager. Copies of the reports are sent to the client.

3. Weekly Utility Reports

The utility coordination on this contract is being handled exclusively by the TTA and the various utility entities that exist in this crowded corridor. The TTA is either letting contracts, as in the case of relocating water and sanitary utilities, or in the case of power and telephone, is paying those companies to relocate their facilities. HNTB's role in respect to utility relocations is twofold. First, through our inspecting techniques, we assure the Authority that the relocated utilities will not conflict with any of the permanent work; and secondly, we are to prepare a weekly utility report, which encompasses the work that has taken place during the past week. These reports include the name of the contractor, the contract number, and the areas where the contractor has performed his work.

4. Miscellaneous Reports

The most important of the miscellaneous reports generated are the monthly payment estimates, whereby HNTB's personnel meet with the contractors on a monthly basis to prepare the payment estimates. Monthly progress reports, and monthly projection of funds required to be paid in the future months (utilized by the Secretary-Treasurer of the TTA) are also prepared. In addition, there are a number of finalization reports prepared to assure the TTA that the contractor has completed his work, and that the contractor has provided a full release of liability, as has any subcontractors and property owners for whom he may have rented property or have potential damage claims against the contractor.

5. Memorandums

All meetings are recorded with memorandums to file. They are accurate records of the meetings and are generally sent to all participants.

In addition, memos to file are written for any accidents on the project, any major involvement with the general public, and the condition of traffic maintenance items.

6. Correspondence

As on any project, correspondence is continually being written between the contractor and the inspecting agency. All matters of importance are covered via letter. All copies of correspondence between the contractor and the construction management team are sent by carbon copy to the contracting agency. If decisions are made that affect the original design, the Design Section Engineer is also sent a copy, especially if his input is required. Other items covered by correspondence are shop drawing transmittals, materials certification, letters in regard to changed conditions, etc.

7. As-Builts

Finally, a complete as-built record plan of the project is required. These records will be placed on drawings identified as "As-Built". A complete set of file drawings is kept in the office. Any changes implemented by the contractor, the Section Engineer, the client, or the GEC are recorded on the as-built drawings. When the project is finally complete, the GEC is charged with completing the transfer of the as-built information to the tracings, leaving the client with a complete set of drawings which reflect as-built conditions.

ANALYSIS AND CONCLUSION

Construction of the Dallas North Tollway Extension began in the late fall of 1983. A conventional construction inspection staff was organized to inspect the construction of the work. As the primary construction contractor revised his staff along specialty lines, HNTB responded by reorganizing along those lines. This staff is presently in place and is doing an excellent job in monitoring the construction of the Tollway. This organization will be continually monitored and revised as needed to insure its appropriateness on contracts recently let, and those to be let in the near future.

The major lesson to be learned from this project is the need for flexibility in order to respond to the contractor's needs. The inspection staff does not construct the project, nor is responsible for supervision of the construction. That is the contractor's responsibility. The inspection staff must be flexible and have a broad base of experience to perform the assigned task in a way mutually beneficial to both the client (in this case the Texas Turnpike Authority) and all contractors.

HNTB will continue to analyze the response time to the contractor's operation. Our organization will be continually monitored and revised as needed to ensure its appropriateness on contracts recently let, and those to be let in the near future. We may return to the conventional resident engineering system, with a resident engineer responsible for each of the contracts. Or, we also may try implementing a hybrid type of system, such as a portion of the inspection organized along product lines and the remainder along contract limits.

This four year project is scheduled to open to traffic by late summer of 1986, with the entire construction project being completed by early winter of 1987.

MANAGING QUALITY INTO THE WORKPLACE

HOWARD A. PEEK AND DONALD J. BROWN*

Introduction

In February, 1983, our organization began a process designed to enhance attitudes about quality in every individual member of the organization. The process requires educating each individual, from Chief Executive Officer to Mail Room Clerk to Laborer, in the principles of Quality Improvement as we now define them. To date, one thousand officers and managers have attended either a 2½ day, or a 4 day training course to learn the principles of managing the Quality Improvement Process. Additionally, almost nine thousand supervisors and employees have attended quality training programs to learn the individual's role in quality improvement. The education effort was predicated on management's conviction that quality, as defined here, must be first among equals with schedule and cost; that is, we must deliver our product or service on time, at the agreed upon price and in conformance to the contract, right, the first time. In order for that to happen, each individual must understand and be committed to quality and to Brown & Root's Quality Improvement Process. We instituted a system of "Quality Management" — a systematic way of guaranteeing that organized activities happen the way they were planned. We formed Quality Improvement Teams at each functional level of the organization to devise methods to improve communications, solicit suggestions for improving our work methods, and investigate and correct situations and conditions that hinder individual employee's ability to do their jobs right the first time. We stress teamwork, recognizing that no one can cause quality improvement alone. We have formed Corrective Action Teams, down to the field crew level, to identify and correct problems as they arise, and to plan preventive measures to keep problems from arising. Traditionally, each individual or operating unit has adopted some abstract definition of quality. We now have only one definition which everyone can understand. The system of quality control has been to detect the defect and fix it. Our system is, ultimately, to prevent the defect from occurring. Formerly the performance standard was a comparison to past performance — "We reworked 10% of the fabricated pipe on the last job, and our rework is only running 8% on this job so we are doing well." Now our performance standard is zero defects —do it right the first time every time. Measurement, too, was a comparison to historical ratios — "Historically, on a project of this type, surplus runs about 9%. Our surplus is running 8½% so we are OK." Now we are calculating the cost of all nonconformances for the purpose of improvement, and applying the standard, Zero Defects.[1] The traditional management approach to quality as opposed to our approach may be seen as follows:

*Howard A. Peek, Construction Manager
Donald J. Brown, Ed.D., Senior Manager, Performance Improvement
Brown & Root, Inc., P. O. Box Three, Houston, Texas 77001

MANAGING QUALITY

TRADITIONAL MANAGEMENT APPROACH		BROWN & ROOT'S APPROACH
Expensive, solid, pretty, shiny, big, bright, etc. (Opinion)	**Definition**	Conformance to agreed requirements the first time every time.
Defect detection and correction. (Fix it)	**System**	Improvement through defect prevention.
That's close enough	**Performance Standard**	Zero defects.
Comparison to historical defect ratios. (Acceptable Quality Levels)	**Measurement**	Cost of nonconformance in dollars

Zero Defects Concept

The concept of Zero Defects, traditionally maligned and misunderstood, does not mean that no one is ever allowed to make a mistake. Rather, it is a sub-system within the quality improvement process. The concept, Zero Defects, is an attitude and a process. The attitude, adopted by each employee is one of commitment. It says, in effect, "I will do my very best to do my job right, in conformance to requirements, the first time, every time." To implement the process, when one does make a mistake, the employee:

1) Defines the situation — What is wrong with this piece of work?

2) Fixes it — make the nonconforming product conform to requirements. This step is designed to keep a defective product from reaching the client.

3) Investigate the defined situation to identify the root cause of the defect. This can be done by oneself, or as a team effort with co-workers, supervisors, vendors, etc.

4) Once the root cause is identified, take action to remove the cause so that that defect never occurs again. It is through the systematic elimination of individual defect causes that we approach the goal of Zero Defects.

No one can cause quality improvement alone. It takes a coordinated effort by all parties to the project; client, contractor, sub-contractors and vendors. Requirements, stated in such a way that they cannot be misunderstood, are solicited from clients, provided to employees and negotiated with vendors.

Supplier Quality

A significant element of the quality improvement process is contractor-vendor relationships. These relationships are developed and maintained by a Supplier Quality management team. Working in concert with key suppliers, teams are established to enhance communication, establish requirements and plan for the elimination of defects in vendor supplied items. The contractor's team is comprised of representatives from procurement, QA/QC, engineering and operations. Regularly scheduled meetings between the contractor and vendor teams identify potential problem areas and plan for problems not to occur. Through these teams we have gained valuable insight in each other's operations and identified problems and constraints we tend to impose on one another. Specific successes to date have resulted from the improved communication systems which are a natural by-product of the quality improvement process.

One vendor had long experienced problems in matching mill test reports (MTR's) with shipments, causing our company extensive record keeping difficulties. Through team negotiations, the vendor initiated a system whereby their internal records do not indicate an order to be complete until mill test reports are received by our project. The MTR's provided traceability of the materials for the owner in the event of defects.

Receipt of MTR's was a client requirement. We were unable to meet this requirement when the supplier did not provide the reports. Consequently, Brown & Root was expending excessive manhours expediting the reports, invoices from the supplier were delayed, and, we were often obliged to request the client to issue waivers and/or change orders.

As a result of the supplier quality team efforts, the supplier began listing all MTR requirements on its order to the producing mill as a line item. The order was not considered complete and payable until the MTR's were received.

When all parties to the process understood and complied with the requirements, the problem ceased, and all parties benefited. Expediting time and cost were reduced, work proceeded on schedule without the need for waivers and change orders, and the vendor was paid in a timely manner.

Another vendor was experiencing difficulties in the timely delivery of critical components to our project. Traditional contractor-vendor interface methods would have eventually solved the problem, but only after considerable delay and increased cost. The supplier quality management teams were able to quickly identify and remove the source of the problem before it severely affected cost and schedule.

Planning

The system, prevention, involves the design of a structured planning methodology to identify potential areas of concern and planning them out of the work process.

On our construction projects, structured planning begins at the project manager level, then cascades downward through the crafts to the crew level. Each level of the project hierarchy understands which functions they have responsibility for planning and each planned activity is coordinated up, down and laterally as appropriate. Nothing is left to chance. Requirements (who, what, where, when and how) are communicated in such a way that they cannot be misunderstood. Potential problem areas are identified and discussed, preventive measures taken and alternate plans devised to deal with problems should they occur. When practical, the crews assigned to do the work walk the area with the foreman or superintendent to familiarize themselves with the work and ask questions or make suggestions. Through planning, support functions such as equipment, tools, scaffolding, etc., are provided with sufficient lead time to enable them to schedule themselves to be where they are needed, when they are needed. A specimen of the planning forms used is included as Attachment "A".

Quality improvement teams, comprised of senior site supervisors, are established for the purpose of managing the on-site quality improvement process. It is common practice for a client representative to attend these team meetings. Recognizing the prevention of problems is preferable to fixing them, these teams assist in developing planning systems to prevent defects and improve productivity. The results have been encouraging.

On one project, drawings for a shut-down were received four working days prior to the shut-down. The client estimated the work would take fourteen, 10-hour days to complete. Through application of the structural planning concept, the work was completed in ten, 10-hour, and three, 8-hour days. More significantly, in testing the work, client inspectors found no defects and did not prepare a punch list.

Another shut-down was estimated to require twenty-two people working fourteen, 10-hour days to complete, and require a 140-ton crane for 1½ days. By applying the structured planning concept, the shutdown was accomplished by seventeen people working seven, 8-hour days and use of the crane for 8 hours. Again, testing was successful and no punch list was required.

These successful shut-downs occurred because of accurate planning, quality workmanship and full client and vendor cooperation.

The impact of the quality improvement process on the quality professional has been significant. Most importantly, it has validated his concept of quality; i.e., that quality is attained when the requirements of the specifications, procedures and codes are met. It has also removed him from the position of arbitrator. No longer must he negotiate from that unenviable position because the concept of "that's close enough" is simply unacceptable. The work in question either conforms to requirements or it does not.

The quality improvement process is a major, permanent, and critical element of our company's long range goals. It began with a firm commitment and dedication to quality by our corporate executives. A corporate quality policy, which reads as follows, has been established and communicated throughout the organization:

> It is the policy of Brown & Root and associated companies that we perform our jobs and deliver services and products in conformance to agreed requirements. Quality shall be first among equals with schedule and cost. Every individual and operational unit shall adopt a standard of performance that demands conformance to requirements and doing things right the first time.

It has been implemented world-wide and involves every aspect of our business.

We are realizing tangible benefits from the process now, but when it fully becomes a way of doing business throughout the organization, we will have created an environment in which the quality professional can truly do his job.

APPENDIX

1. Crosby, Philip B.
   <u>Quality is Free</u>. (New York, New York, New American Library, Inc.)

PIPE PLANNING FORMS

PIPE SUPERINTENDENTS MONTHLY SCHEDULE							
SYSTEM	ISO	DATE ISSUE MAT. CONTROL NEEDED	LINEAR FEET	GENERAL FOREMAN	DATE COMP.	COMMENTS	

PIPE GENERAL FOREMAN—TWO WEEK SCHEDULE						
SYSTEM	ISO	MAT. AVAIL. SPOOL/ERECTION	SPOOL SEQUENCE	DATE FOREMAN	COMP.	COMMENTS

PIPE FOREMAN—TWO WEEK SCHEDULE							
ISO SPOOL	REQUISITION NO. DATE ISSUE	NEED DATE MAT. REC.	EQUIP. REQ.	SCAF. REQ.	FIELD FAB.	EST. M/H	ACTUAL COMMENTS

Attachment A

SUBJECT INDEX
Page number refers to first page of paper.

Case reports, 1, 26
Construction management, 5, 26

Highway construction, 26

Management methods, 15
Management training, 38

Organizational policy, 38

Performance standards, 15
Productivity, 5
Project managers, 21

Quality assurance, 1, 21
Quality control, 1, 5, 15, 38

Teamwork, 38
Toll roads, 26

U.S. Army Corps of Engineers, 15

AUTHOR INDEX
Page number refers to first page of paper.

Brown, Donald J., 38

Johnstone, Homer, 15

Little, Paul, 21

Peek, Howard A., 38

Quinn, Stephen B., 26

Stukhart, George, 1

Tatum, C. B., 5